RAINTRE
SCIENCE ADVENTURES

THE MOON

Helen H. Carey and Judith E. Greenberg
Illustrated by Donna Corvi

Raintree Publishers
Milwaukee

To Mary Ellen and Fran — **H.C.**
To Aunt Alice — **J.G.**

Editorial

Barbara J. Behm, Project Editor
Judith Smart, Editor-in-Chief

Art/Production

Suzanne Beck, Art Director
Kathleen A. Hartnett, Designer
Carole Kramer, Designer
Andrew Rupniewski, Production Manager
Eileen Rickey, Typesetter

Reviewed for accuracy by:

Gretchen M. Alexander, Executive Director
West 40 Educational Service Center Number 5
Northlake, Illinois

Scott R. Welty, Instructor
Maine East High School
Park Ridge, Illinois

Library of Congress Number: 89-78291

1   2   3   4   5   6   7   8   9   94   93   92   91   90

Library of Congress Cataloging-in-Publication Data

Carey, Helen H.
   The moon / by Helen H. Carey and Judith E. Greenberg;
illustrated by Donna Corvi.
   (Raintree science adventures)
   Summary: Describes the surface and movement of the moon, eclipses, the moon's effect on tides, and the possible origin and formation of the moon.   1. Moon—Juvenile literature.
[1. Moon.]   I. Greenberg, Judith E.   II. Corvi, Donna, ill.   III. Title.   IV. Series.

QB582.C37   1990   523.3   89-78291
ISBN 0-8172-3752-6 (lib. bdg.)

Before You Begin

This book takes *you* on an adventure to the moon! You will travel over 200,000 miles (321,800 kilometers) to reach the moon from the earth. The moon is one-fourth as big as the earth.

You will find out what it's like to be on the moon, how the moon affects the ocean tides on earth, and how the moon may have been formed. Your adventure will include two science experiments.

Knowing the words below will help you in your adventure.

axis   a real or imaginary line that passes through an object, on which the object turns

crater   a bowl-shaped hole, such as those found on the moon's surface

eclipse   the darkening of a body in space, such as the sun or the moon, when the shadow of an object in space falls on it or when an object moves in front of it, blocking the light

gravity   the natural force that causes objects to move toward the center of the earth or the moon

orbit   the path of one object around another object in space

phase   a stage, such as the different stages of the moon at different times of the month

reflect   to turn or throw back light, heat, sound, and so on

revolve   to move in a circle around something

rotate   to turn on a center point or axis

Now turn the page, and begin your science adventure!

On the Moon

It is close to eight o'clock in the evening. Tomorrow you are going to give a report about the moon to your class. You are looking through a science magazine that just came in the mail. You hope to find a picture of the moon to use with your report. Finally, on the last page of the magazine, you see THE PICTURE.

It is a picture of a moon station. The story with the picture says that someday soon, people will live in this moon station. They will work in moon mines. They will send moon minerals back to earth. You wish that you could live in a moon station.

The story says that people are welcome to come to the new science center in your city to see a model of the moon station. Real astronauts will be there explaining how the moon station will work.

The next day, you are the first one to give your report in class. When your classmates look at the picture that you found, they all want to visit the model of the moon station.

"This is a good time to show all of you my special surprise," says your teacher, Ms. Kornspan. She holds up a small, clear plastic box with something inside it. She passes it around. Everyone tries to guess what is inside the little box.

"I think it's a pebble from your driveway," says Ricky.

"I think it's a piece of candy from outer space," says Sarah.

"Sarah and Ricky are both partly right. It is a rock, and it is from space. Now who knows exactly what it is?" asks Ms. Kornspan.

"I know! I know!" you say. "It's a moon rock!"

Ms. Kornspan smiles and puts the moon rock on your desk.

"Yes, that's correct. Astronauts brought it back from the moon. Today, I want each of you to look at the moon rock. Then imagine that you are in a spaceship that has just landed on the moon. Write on a piece of paper what you think you will see on the moon," Ms. Kornspan says.

You close your eyes and think about landing on the moon. After a while, you become sleepy. You begin to feel kind of funny. You feel light as a feather. You are out of the pull of the earth's **gravity.** There is no up or down for you. You hold on to your chair. Otherwise, you will float away.

You are traveling to the moon in a spaceship. You hear the ship's engines start working to slow the ship down. A great cloud of dust blows up around the ship as it lands. You have landed on the moon!

Before you go outside the spaceship, you put on your special boots and gloves. Then you put on your backpack that contains a supply of air. The air supply is necessary because there is no air or oxygen on the moon. On earth, your equipment would weigh 150 pounds (68 kilograms). On the moon, it weighs only 25 pounds (11 kg)! That is because the force of gravity on the moon is one-sixth of that on earth.

Outside the spaceship, many things are very different than on earth because there is less gravity. With just one step, you move ten feet forward. You kick the rubber ball that you've brought along. It travels six times as far as it would on earth.

In darkness, you look at the sky above the moon. It is very black. You can see many more stars than you can see from earth. On the moon, you do not have to look through layers of air and pollution. Therefore, the stars seem brighter.

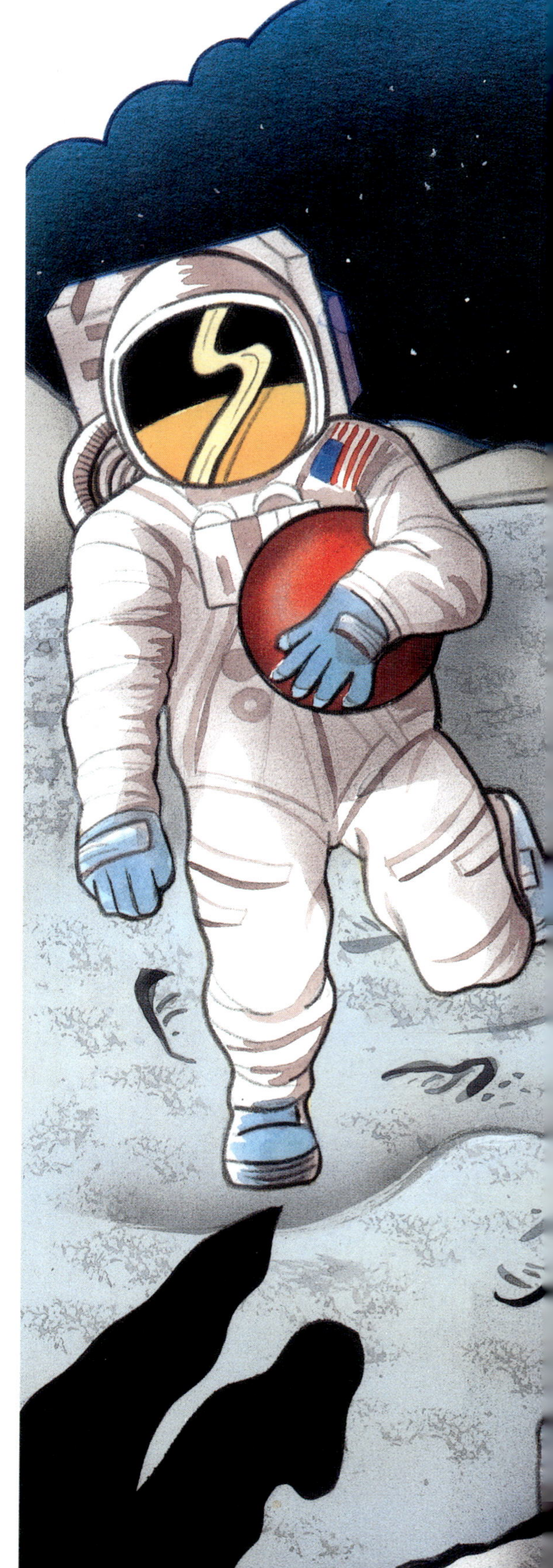

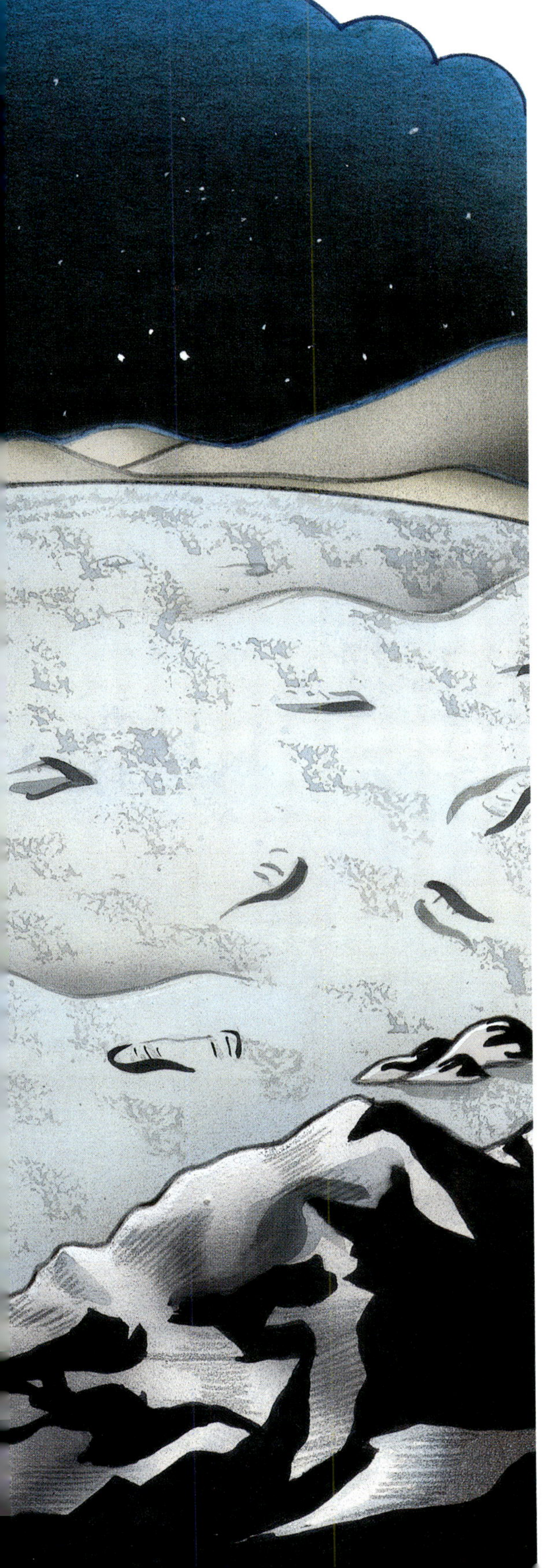

It never rains or snows on the moon. In shadows and darkness, the temperature is freezing cold. In sunlight, the temperature is burning hot. However, in sunlight, the sky is not blue like on earth. Instead, the gold of the sun fills the sky.

On the moon, you can yell and try to make all the noise you want, but there is only silence. This is because there is no air to carry sound waves.

You use a scoop to pick up some moon rocks. The rocks are grayish brown and look like the rocks at home.

Home? Mmmmmmm . . . .

The next thing you know, someone is tapping you on the shoulder. You are back in your classroom.

"Come on! Let me see the moon rock. You've had it so long, I thought you went to the moon!" says Sarah.

The Changing Moon

The next day, you go on a field trip with your class to the science center. The floor of one of the rooms at the science center was built to look almost exactly like the surface of the moon. It is bumpy and full of big and little holes. The holes are supposed to be **craters.** You look up and see an exhibit of the moon's sky in darkness. It is inky black with twinkling stars. You see the model of the moon station. For a moment, you think you are back in your dream.

Pictures of the earth as it looks from the moon cover the wall.

The surface of the moon has lowlands and highlands. From earth, the lowlands on the moon look like dark patches. The bright parts that you see on the moon are the highlands.

You want to know why the moon shines. You pick up a telephone that is part of the exhibit. A recorded voice tells you that the moon makes no light or heat of its own. Instead, the moon is like a giant mirror. The moon **reflects** some of the bright light from the sun. This reflected light makes the moon glow in the sky. You can only see the moon by the light it reflects from the sun.

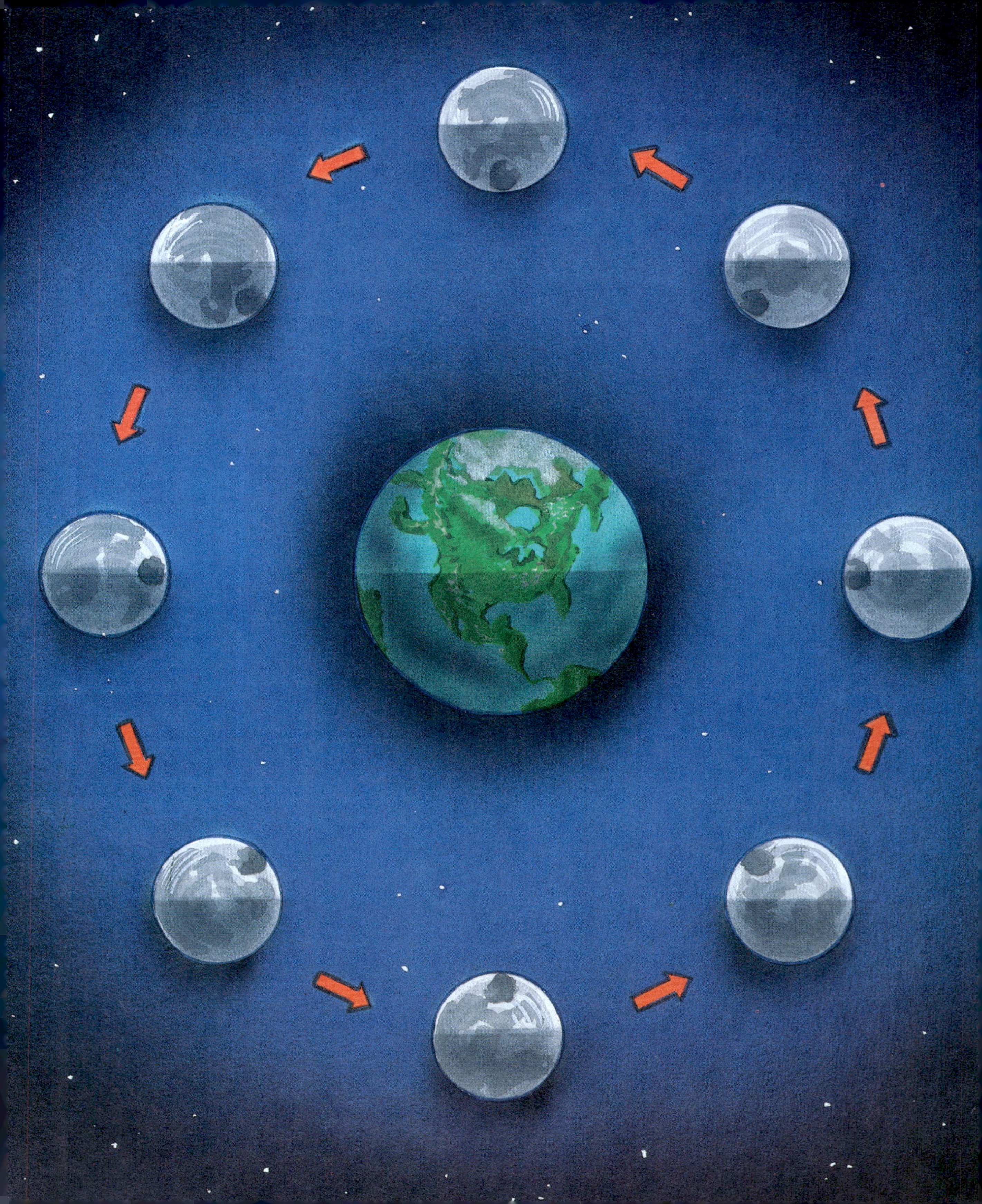

You take an elevator to the lower level of the science center. In this exhibit hall, pictures of the moon cover the entire length of a big room.

In each picture, you see the lighted side of the moon. This is the side that always faces the earth. You wonder out loud, "Why does only one side of the moon ever face the earth?"

A man standing next to you hears your question. He says, "Just as the earth does, the moon **rotates** on its **axis.** The moon rotates once every twenty-eight days. It also takes the moon twenty-eight days to circle, or **revolve** around, the earth.

"In other words, the moon turns once on its axis in the same amount of time that it takes it to circle the earth. Therefore, we always see the same side of the moon."

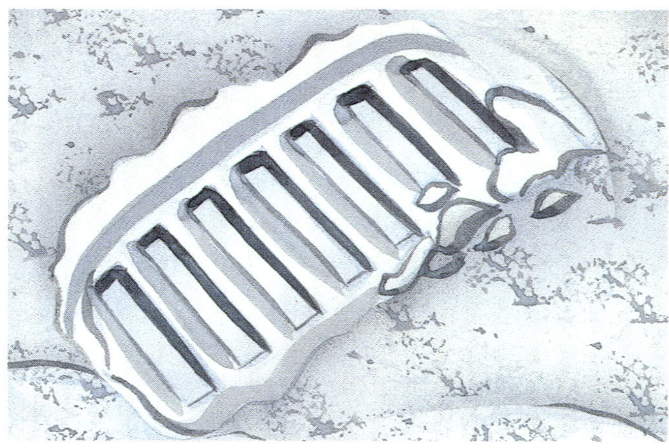

"Gee, how do you know so much about the moon?" you ask the man.

"Well, the moon is part of my job," the man says with a smile. "I'm an astronaut. I have even walked on the moon."

You can hardly believe your ears!

"My science book has a picture of an astronaut's footprint on the moon. Is it yours?" you ask your astronaut friend.

"It could be," the astronaut says. He leads you across the room to see some of the other moon pictures.

You ask the astronaut, "Why does the moon change shape? Some nights I see the whole moon. On other nights, I just see half of it. Sometimes it looks like the end of my fingernail."

"The moon doesn't really change shape," replies the astronaut. "What happens is that it goes through **phases,** or stages. It does this as it revolves around the earth throughout each month.

"The *new moon* is the first phase. In this phase, you can't see the moon at all. That's because the moon is between the earth and the sun. The sun shines on the side of the moon that is turned away from the earth. You cannot see any of the lighted side of the moon. You just see the dark sky.

"The next major phase is the *waxing crescent* phase. *Waxing* means that something is growing bigger. During this phase, the moon moves a little bit from its position between the sun and the earth. You can begin to see a part of the lighted side of the moon. The curved shape you see is this part.

"Little by little, you see more and more of the lighted side. When you see half of the side of the moon that faces earth, about seven days after the new moon, you see the phase called the *first quarter*. It is also known as the *half moon*.

"In another seven days or so, is the *full moon*. Now you can see the whole side of the moon that faces earth.

"The *third quarter* of the moon comes about seven days later. The moon looks like a half moon again.

"Next is the *waning* crescent. *Waning* means that something is growing smaller.

"A week after the third quarter, the cycle begins again with a *new moon*," the astronaut says.

It was fun to meet a real astronaut. You thank him and say good-bye before you continue on your science adventure.

New moon

Waxing crescent

First quarter

Full moon

Third quarter

Waning crescent

Eclipse!

You enter a small room of the science center. You look around and see a sign that says LUNAR ECLIPSE EXPERIMENT.

A guide is standing beside the sign. You walk over and ask, "May I do this experiment?"

"Certainly," answers the guide. "I can help you." She hands you a flashlight, a Ping-Pong ball, a light blue balloon, and a felt-tip pen.

The guide asks you to take the balloon and blow it up. Then she tells you to draw shapes of the earth's continents on it with the felt-tip pen. Now the balloon looks something like a globe. You hold the balloon up and shine the flashlight on it. You see the balloon's shadow on the wall behind the balloon.

The guide says, "Imagine that the balloon is the earth. The flashlight is the sun. Pretend that this Ping-Pong ball is the moon. Now watch." She moves the Ping-Pong ball behind the balloon and into the balloon's shadow. "What do you think will happen to the Ping-Pong ball when it is in the shadow of the balloon?"

(Write your guess on a piece of paper.)

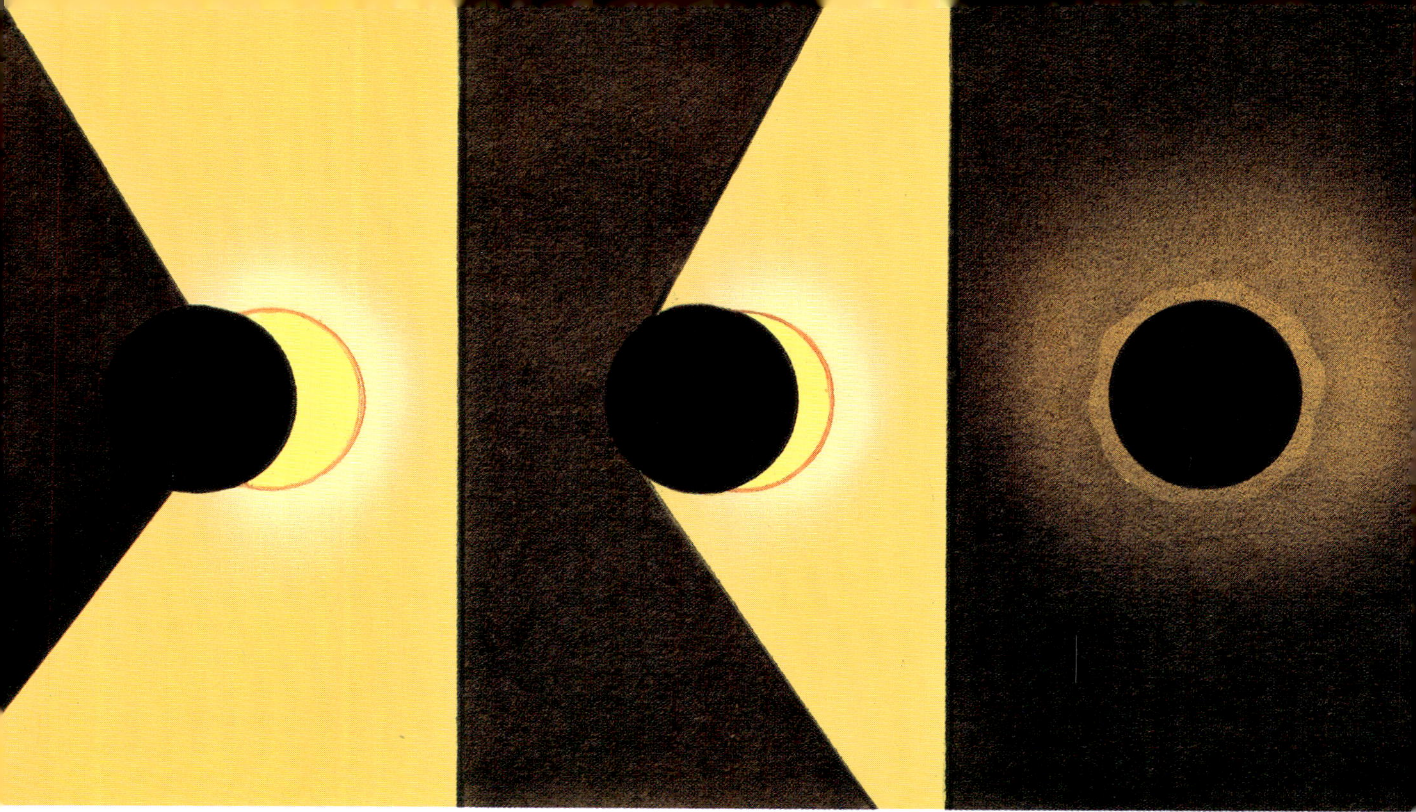

Solar eclipse

"The Ping-Pong ball stays dark," the guide says. "It does not reflect the light because it is in the balloon's shadow. This is an example of a lunar **eclipse.**

"In a real lunar eclipse, we see the moon get darker and darker as it moves into the shadow of the earth. The earth's shadow covers the moon. Then, gradually, the moon moves out of the earth's shadow, and we see it glowing again."

"An eclipse of the sun, which is called a solar eclipse, is exciting, too. That's when the shadow of the moon passes across the earth," says the guide.

"How does a solar eclipse happen?" you ask.

"A solar eclipse takes place when the moon slowly passes between the sun and the earth and blocks out the sun. The sky darkens, and the birds think it's night! After a few minutes, the sky gradually begins to get light again, and the birds begin to sing," the guide says.

She tells you never to look right at the sun during a solar eclipse or at any other time. Even with sunglasses, your eyes can be hurt at any time by the rays of the sun. Scientists use telescopes with special cameras to take pictures of a solar eclipse.

Danger in the Tide!

On Sunday, you go to the beach with your mother; your younger sister, Susan; and your dog, Piper. Piper chases anything you throw and brings it back to you. You throw a ball into the water. You play fetch with Piper for a long time.

"Come up here for a while," calls your mother. "You and Piper need to rest." You race Piper to the beach umbrella.

"I'm going to the refreshment stand to get us some lunch," your mother says to you. "I'll be back soon. Stay out of the water, and watch Susan while I'm gone. If there's any problem, ask the lifeguard over there for help."

Your sister is asleep on the blanket, so you build a sand castle. You put a road around the castle walls.

Piper is barking. You are busy finishing the road, so you pay no attention. He runs and jumps, barking louder and louder. You look toward Susan's blanket. She is gone. The ocean water has come up to the blanket and soaked it.

You see your sister crawling into the water. The next wave will crash over her head. Neither you nor the lifeguard can reach her in time, but Piper can! "Fetch, Piper! Fetch Susan!" you yell.

Your mother rushes back just as Piper pulls Susan out of the water. The lifeguard makes sure Susan is all right. Then Susan, you, and Piper are grabbed up in your mother's big hug. You are all glad that Piper came to the rescue.

Everyone sits down to recover from the scare.

"The water came up to Susan's blanket because the tide came in. The tide is the regular rise and fall of the ocean," says your mother. "When we first came to the beach today, it was low tide. The water was at a low level. Much of the beach could be seen. Now it is high tide. The water is at a high level. Less beach can be seen. There are two high tides and two low tides each day."

You want to know what causes the tides.

"The moon is the main reason there are tides," your mother answers.

She continues, "The moon's gravity pulls up the water on earth that is located directly below the moon. This causes the water there to gradually rise for six hours, forming a high tide. At the same time, on the other side of the world, there is also high tide. This happens because the moon pulls the solid earth away from the water.

"As the earth turns, the tides rise and fall. After high tide, the waters begin to fall for a period of six hours. When the water level is at its lowest point, it is called low tide. The cycle then begins again until high tide is reached."

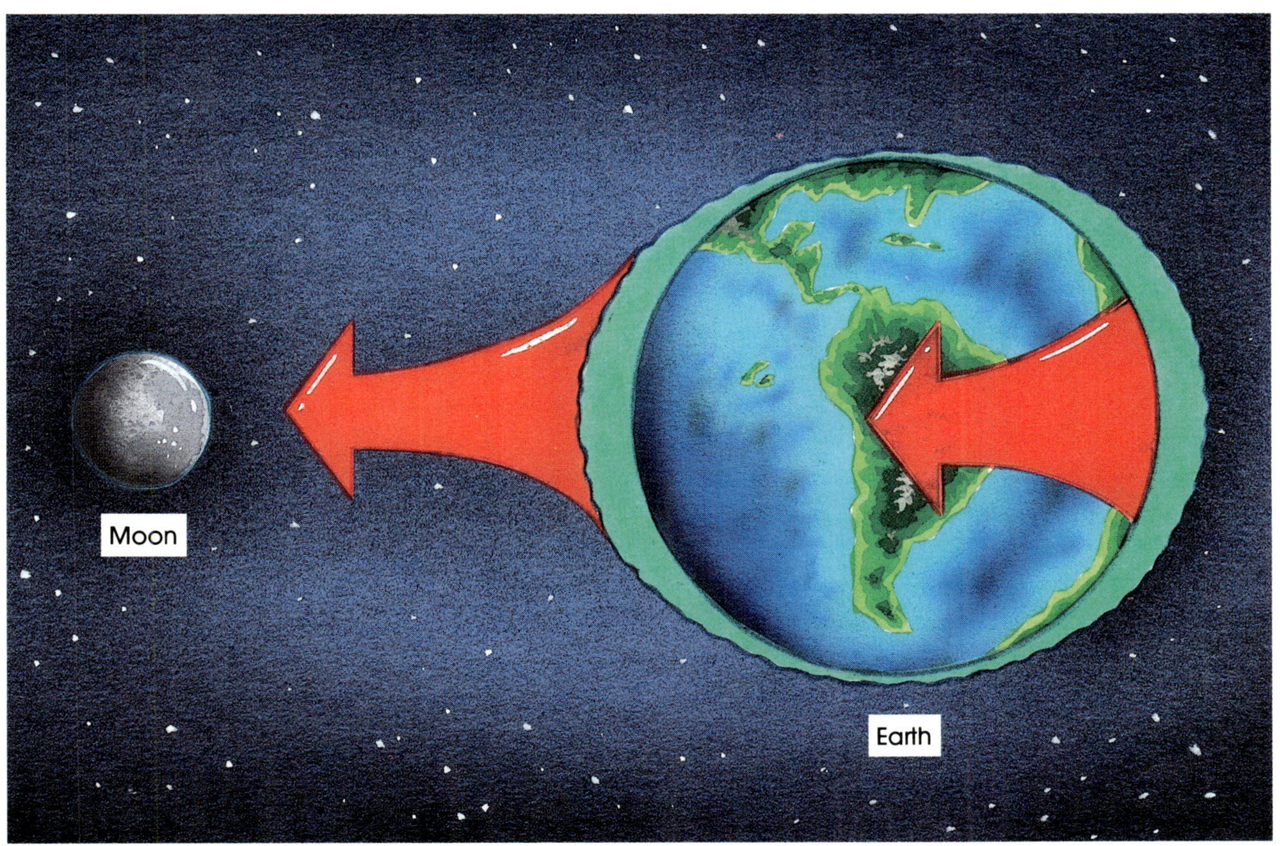

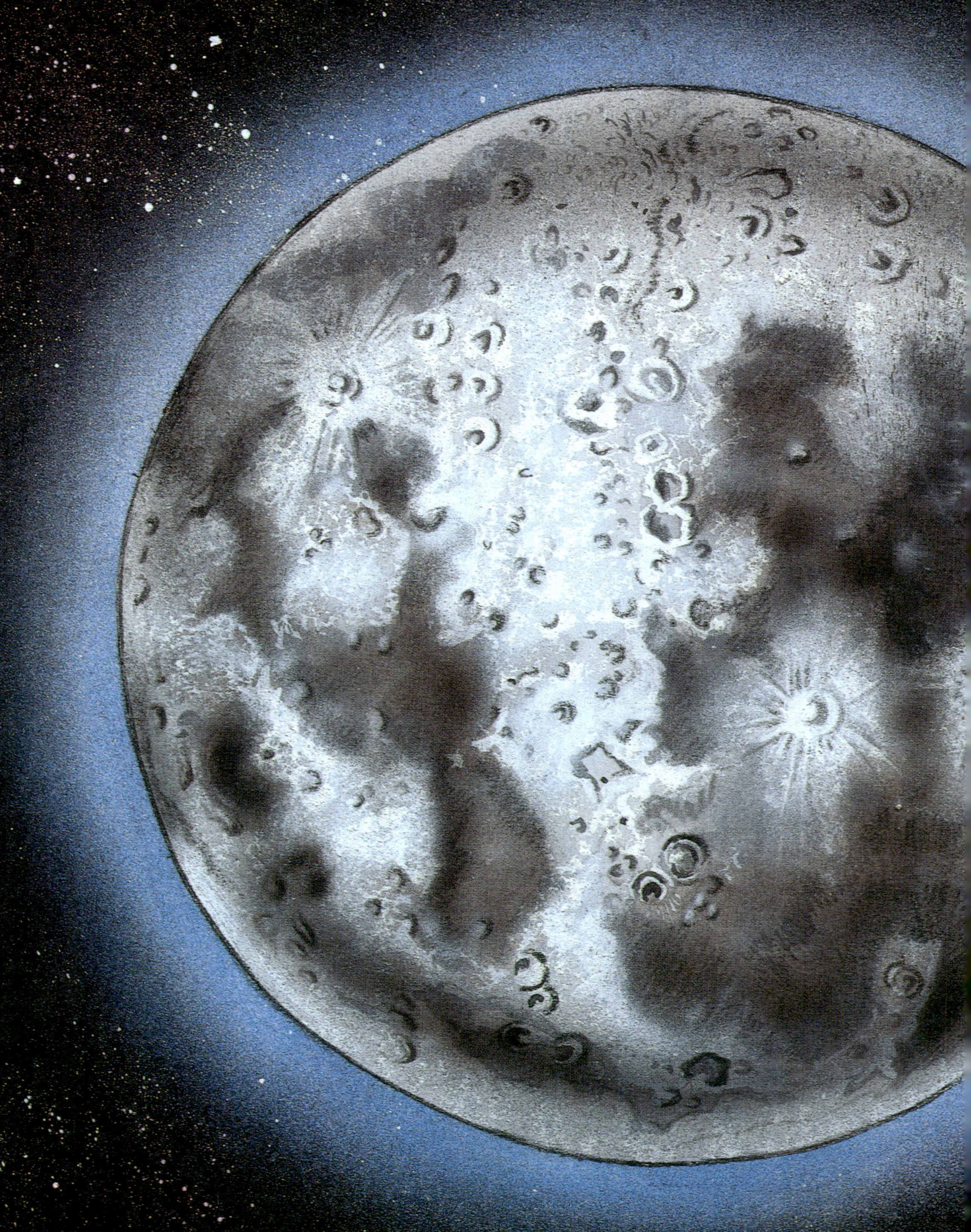

Where the Moon Came From

On Monday, you are back in school.

"Good morning, class," says Ms. Kornspan. "I invited an astronomer to class today. An astronomer is a scientist who studies the planets and stars. Meet Dr. Donovan."

Everyone wants to ask Dr. Donovan a question. "Why does the moon sometimes have a ring around it?" asks Sarah.

Dr. Donovan answers, "Sometimes the sky contains very high clouds that are made of tiny ice crystals. When the light from the moon shines on these clouds, the light scatters. From the earth, this light appears as a glowing ring around the moon."

"Why are there so many craters on the moon?" asks Ricky.

"There are so many craters because there are so many meteorites. Craters are formed when meteorites, which are pieces of solid matter in the solar system, crash into the moon," answers Dr. Donovan.

You raise your hand. "Where did the moon come from?"

"Scientists do not know for certain how the moon was formed. However, they have come up with different theories, or beliefs, about what may have happened. These pictures will help explain," answers Dr. Donovan.

Escape Theory

The moon was once part of the earth. At some point, the moon broke off or "escaped" from the rapidly spinning earth and became a separate body.

Double Planet Theory

The earth and moon were formed at the same time in the same orbits they are in today. They were probably formed from the dust and gases that were left in space after the sun was formed.

Capture Theory

The moon was a planet that had its own **orbit** around the sun, just as the earth does. The moon kept coming closer and closer to the earth. One time when the earth and moon passed each other, the earth's gravity pulled or "captured" the moon. This capture put the moon in orbit around the earth.

Crater Experiment

Ms. Kornspan's class invites Dr. Donovan to come back for the class science fair next week. Everyone is excited. All the projects will be about the moon.

Ricky is making a model of a moon station. He is building a moon car out of pipe cleaners with buttons for the wheels. Sarah is making a poster of the ring around the moon. You are making craters like the ones on the moon.

You fill a tray with moist dirt. You find different sizes of rocks to use as meteorites.

You drop a rock as big as your fist onto the dirt. You make a giant crater! You toss a few of your other smaller rocks onto the dirt, too. Sometimes you stand right over the dirt. Sometimes you throw from the side or from a few feet away. You take each rock out of the dirt before you throw the next rock.

With your hands, you mold peaks onto some of the craters. You make craters with smaller craters inside of them. Soon the dirt looks like the surface of the moon. You use a camera to take a picture of it.

You compare the photograph of your moon craters to one that you find in a book.

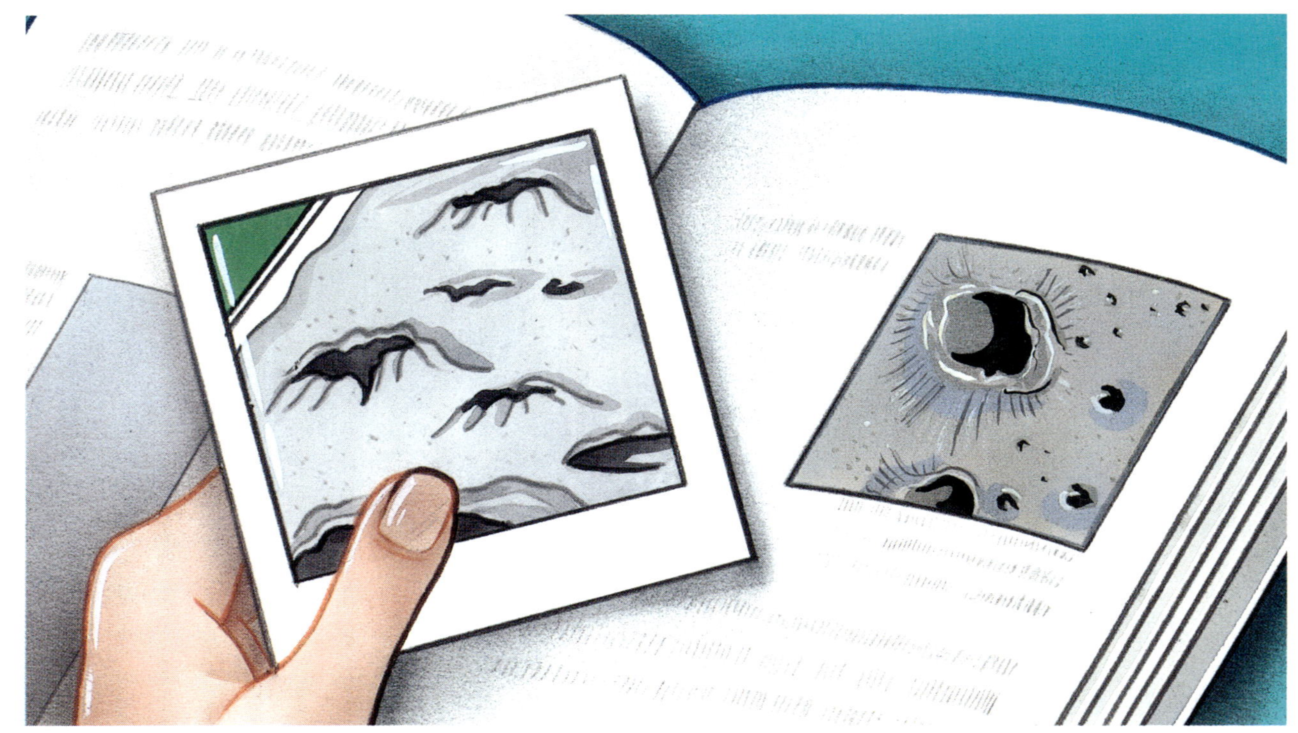

During the science fair, you talk about your
experiment to your classmates. You say that craters are
bigger than the meteorites that formed them. The size
and shape of a crater comes from material being
exploded out of the place where the meteorite hits. Big
and small meteorites hit the moon from all different
directions, making the craters. You explain that you
tested what you learned about craters by throwing
different-sized rocks from different directions at the dirt
in a tray. Everyone is interested in the craters.

You have liked learning all about the moon. You
might someday be a scientist who studies the moon.
Perhaps you will travel there one day.

What kind of science adventure would you like to go
on next?